MICROBES
THE TINY CREATURES AROUND US

Written by Rebecca Borger • Designed by Robin Fight

© 2022 Jenny Phillips | goodandbeautiful.com

To Nathanael, the one whose wonder amplifies mine.

Psalm 147:4–5

There are tiny creatures all around us—

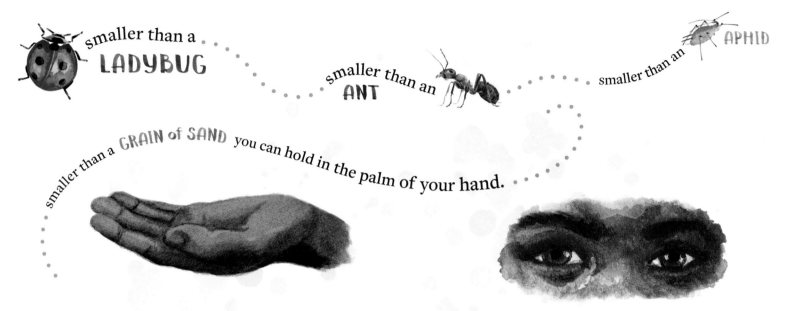

smaller than a **LADYBUG**

smaller than an **ANT**

smaller than an **APHID**

smaller than a GRAIN of SAND you can hold in the palm of your hand.

Sometimes they are so small you cannot see them with your own two eyes.

They can be so small that more than 200,000 of them would fit on the period at the end of this sentence.

200,000

Two hundred thousand!

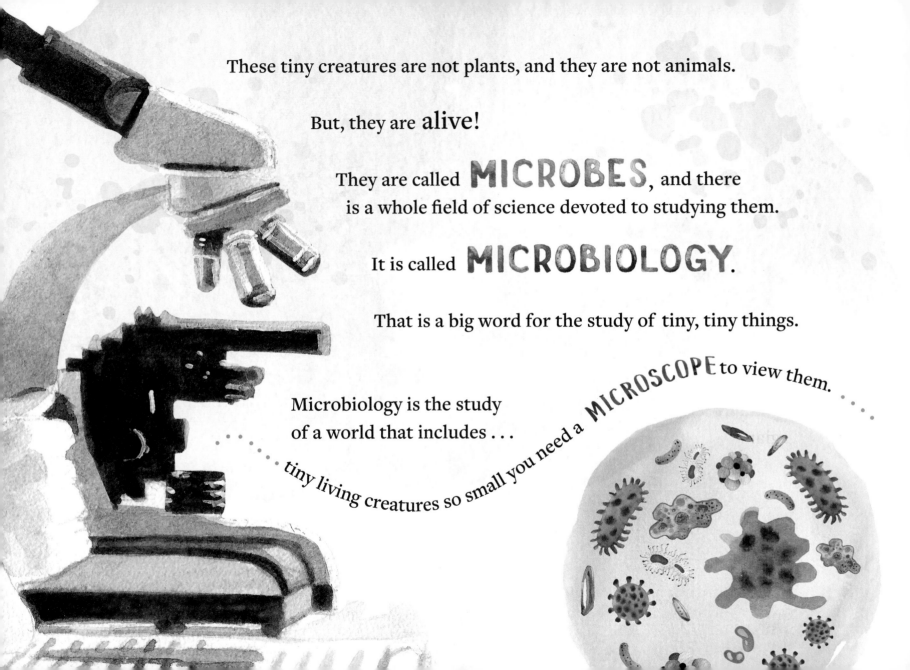

These tiny creatures are not plants, and they are not animals.

But, they are **alive!**

They are called MICROBES, and there is a whole field of science devoted to studying them.

It is called MICROBIOLOGY.

That is a big word for the study of tiny, tiny things.

Microbiology is the study of a world that includes . . . tiny living creatures so small you need a MICROSCOPE to view them.

Microbiology is the study of microbes all around us, everywhere.

This world of tiny creatures is full of big microbes,

like an AMOEBA that swims in the ocean,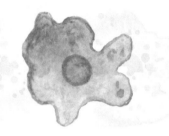

and small microbes,

like BACTERIA deep in the soil,

just like the world you can see with your own two eyes is full of

big creatures, like a giant elephant,

and small creatures, like a baby mouse,

but even the biggest microbe is small to you.

Microbes live **everywhere.**

They live where there are people, plants, and animals—

and they live where no people, plants, or animals can ever live.

You can find them up in space
(at the International Space Station) . . .

and in the deepest part of the ocean (in the Mariana Trench).

You can find them at national parks
(in the Morning Glory Rainbow Pool
at Yellowstone National Park) . . .

and in the dirt in your own backyard garden.

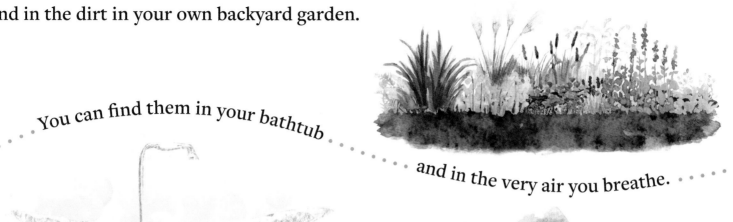

You can find them in your bathtub

and in the very air you breathe.

You can find them in extreme **HEAT**,
and you can find them in extreme **COLD**. Microbes that live
in environments so hostile no other life can survive are called **EXTREMOPHILES**.

Most interesting of all, you can find microbes, tiny creatures of the unseen world, on the outside and the inside of you!

This is called the **HUMAN MICROBIOME**, and there are at least as many microbes in and on your body as there are your very own cells! Some scientists say there are ten times as many.

The human microbiome is the name for all the microbes that live in harmony on and in the human body.

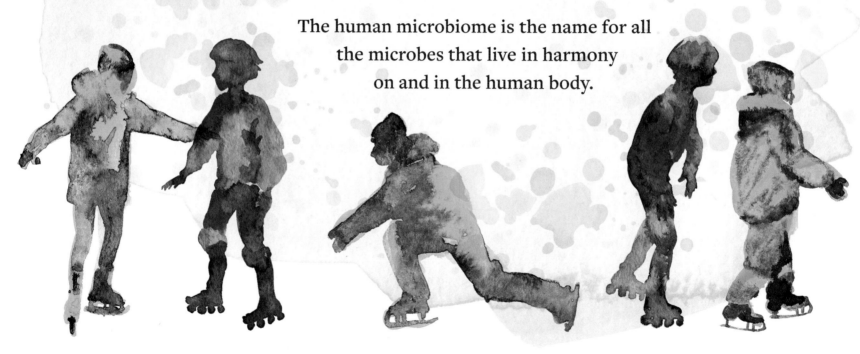

These microbes work to help your body do all the wonderful, marvelous things it can do.

In every place microbes are found, tiny creatures are doing giant things.

Microbes make the soil smell fresh and sweet.
They make the earth healthy and good.

Microbes break down big things into small things that can be used again and again,

like helping turn rocks into soil

or fallen leaves into mulch.

They are the best recyclers in the world!

Microbes can turn the ocean red (this is called a RED TIDE),

or make a pond green (this is called an ALGAL BLOOM).

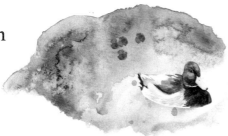

Microbes help clean up oil spills, and they help make lifesaving medicines.

Microbes help your body stay balanced and healthy.

Most microbes help, but some can and do cause harm.

Microbes that cause harm are called

PATHOGENS or GERMS.

Have you ever had a runny nose, a fever, or an oozy sore?
A microbe is to blame!

Some microbes cause infection, but the majority of microbes are helpers.

There are ways to keep out harmful microbes.
Here are some ways to keep out bad microbes and help the good ones.

 Wash your hands.

 Cover your mouth when you cough and sneeze.

Keep surfaces clean.

Be careful around people when you or they are sick.

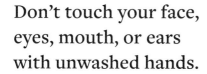 Follow food safety practices.

 Protect broken skin, like a cut, from dirt and debris.

Don't touch your face, eyes, mouth, or ears with unwashed hands.

If you are not sure what is needed to keep harmful microbes out, you can always ask a parent or guardian to help you!

It is good to have helpful microbes doing their jobs . . .

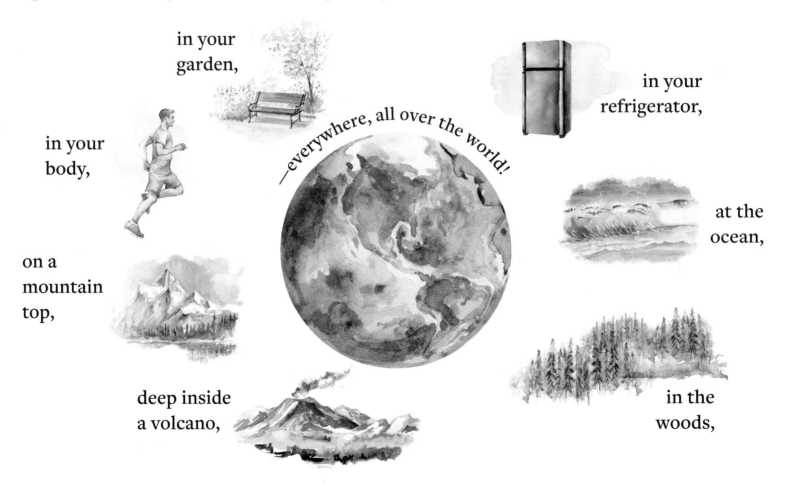

in your garden,

in your refrigerator,

in your body,

—everywhere, all over the world!

at the ocean,

on a mountain top,

deep inside a volcano,

in the woods,

When it is clear and dark, step outside. Look up at the velvet black sky.

Stars shimmer and wink above your head, stretching on and on and on.

Some appear large and bright; others appear small and faint.

There are stars that are invisible to your own two eyes, even with a telescope.

Can you count all the stars?

Scientists believe that there are more microbes in our world than there are stars in the sky!

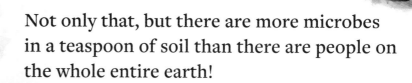

Not only that, but there are more microbes in a teaspoon of soil than there are people on the whole entire earth!

Microbes are numerous and diverse, and many still haven't even been discovered yet!

Even though microbes are so tiny they are not all the same.

... These tiny creatures are varied and intricate and UNIQUE.

Microbes are beautiful, interesting, and sometimes STRANGE.

A microbe can look like ...

a corkscrew, or ...

a fuzzy circle, or ...

a spider, or ...

a snowflake, or ...

a tube.

If you study these pictures of microbes, what shapes and designs do you see?

Microbes come in many different shapes, designs, and sizes, just like the creatures of the world you are used to seeing.

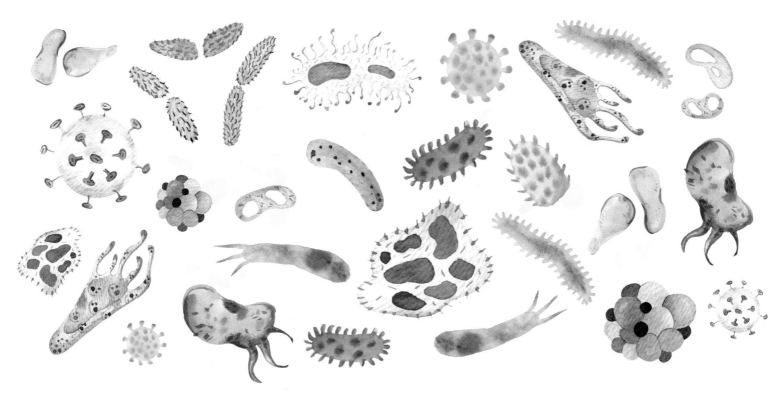

Think of the variety you can see every day when you look around you!
There is an invisible world full of just as much variety all around you, too.

Microbes can be divided into different categories such as bacteria, archaea, viruses, fungi, protozoa, and algae.

BACTERIA

can be very helpful, but some are harmful. Bacteria help cycle nutrients in the soil, help create healthy foods for us to eat, and can also cause food to spoil.

LACTOBACILLI   CLOSTRIDIUM BOTULINUM   STREPTOMYCES

ARCHAEA

are often found living in extreme environments such as hot springs with temperatures over 100 degrees Celsius (212 degrees Fahrenheit) or in the frigid ice of the Arctic Ocean. These are the extremophile microbes, living where other life cannot survive.

METHANOPYRUS KANDLERI   HALOBACTERIUM SALINARUM

VIRUSES are the smallest of all microbes and cause illnesses, including these:

COMMON COLD — INFLUENZA — CORONAVIRUS

Viruses are unique among microbes because they are not considered true living creatures. Viruses cannot do anything on their own; they need a HOST

(a human, plant, or animal, for example) in order to do anything.

FUNGI

While some types of fungi are large and visible to the eye (mushrooms and some kinds of mold), many others are microscopic and part of the invisible world. This is why fungi are included in the world of microbes. Fungi act as decomposers, help create foods such as cheese and chocolate, and are also a source of antibiotics!

PENICILLIN

ASPERGILLUS

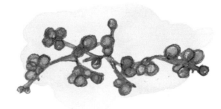
CANDIDA

PROTOZOA

are tiny mobile living creatures. They are bigger than bacteria and love moist environments. Many live in the ocean, ponds, and lakes. Some live in soil.

PARAMECIUM

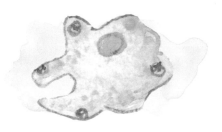
AMOEBA

ALGAE

Even though some microbes are invisible to the naked eye, these tiny microbes can sometimes grow together into large groups or chains that you can see. This is definitely true for the microbe algae! Have you ever seen a bright green coating floating on the surface of a pond? If so, you are looking at a huge group of microbes. Most algae grow in the ocean and seas or in bodies of fresh water, but some grow on rocks and trees. Some even grow in an animal's fur, like in the fur of a sloth or polar bear!

DIATOMS   CHLORELLA

Microbes are the tiny living creatures all around us.

But they don't eat like we eat, and they don't breathe like we breathe. And they **MULTIPLY** (make more of themselves) differently, too.

Microbes are found everywhere in the world, and everywhere they are found, they are eating. They can eat ...

soil, plants, rust, rock, and even oil.

And everywhere they are eating, they are transforming one thing into something else.

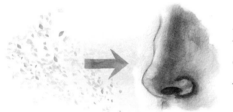

Microbes can turn chemicals in the air into the oxygen you breathe.

They can turn milk into yogurt.

Microbes can turn vegetable scraps into garden compost.

They can turn metal into fine, rust-colored powder.

Microbes feed on **anything!**

Most microbes multiply very rapidly. They reproduce or replicate much faster than . . .

humans, plants, and animals.

A microbe replicates by splitting apart and making more of itself. So, if there is one microbe, it will split apart and then there will be two. Shortly after that, those two will split, and then there will be four, and then those four will split, and there will be eight. Then 16 . . . and on and on!

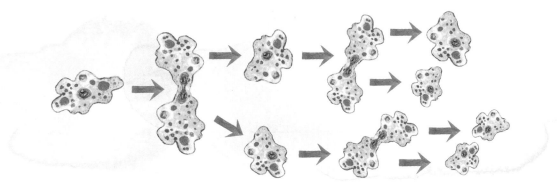

Until, before you know it, in 24 hours, there can be more than a trillion new microbes.

Now you can see why this microscopic world has more tiny living creatures than there are stars in the sky!

Microbes are the tiny creatures around us that help keep our world healthy.

They are on the outside and the inside of you and me and help our bodies do wonderful things like digest food and fight disease.

Microbes help form delightful and healthful things like bread and cheese,

and they can cause illnesses like strep throat or an ear infection.

A microbe may have helped form the antibiotic that helps you get better from that strep throat or ear infection!

Microbes are good helpers, but some do cause harm.
They are tiny active workers doing big and important jobs.

When you wake up tomorrow and open your own two eyes, look around at everything you can see.

Remember, there is an unseen world—full of tiny creatures,

smaller than a BEETLE.... smaller than a HONEY BEE.... smaller than a GRAIN of RICE....

smaller than a FLEA.

These tiny living creatures are doing the hard work of recycling, reusing, restoring, and transforming everywhere you can think of, all over the world.